Henry Saffory

The Inefficacy of All Mercurial Preparations in the Cure of Venereal and Scorbutic Disorders,

proved from reason and experience. With a dissertation on Mr. de Velnos's vegetable syrup, which radically cures every species of the above disorders

Henry Saffory

The Inefficacy of All Mercurial Preparations in the Cure of Venereal and Scorbutic Disorders,
proved from reason and experience. With a dissertation on Mr. de Velnos's vegetable syrup, which radically cures every species of the above disorders

ISBN/EAN: 9783337310905

Printed in Europe, USA, Canada, Australia, Japan

Cover: Foto ©berggeist007 / pixelio.de

More available books at **www.hansebooks.com**

TO THE

MEMBERS

OF THE

ROYAL COLLEGE of PHYSICIANS;

THIS

SMALL TREATISE

ON THE

INEFFICACY OF MERCURY

IN THE

CURE OF VENEREAL

AND

SCORBUTIC DISORDERS,

Is humbly inscribed,

By their Most Obedient Servant,

HENRY SAFFORY.

PREFACE.

EVERY member of Society is under an indiſpenſable obligation to promote, as far as he is able, the happineſs of the whole community. The ſoft whiſpers of humanity plead ſtrongly in behalf of the diſtreſſed; and the heart of that man muſt be callous indeed, who withholds relief from the cries of anguiſh, when that relief is in his own power. The human frame is ſubject to many diſorders: pain and diſeaſe are part of the portion of morta-

A   lity.

lity.   Providence has indeed amply fur-
nished the earth with medicines for af-
fuaging the one, and removing the
other ; but the knowledge of these fimples, their virtues, and the manner of
application, are left to the fagacity of
mankind ; a task not eafily performed.
They are scattered in such amazing pro-
fufion, and in such aftonifhing variety,
that the mind is perplexed ; and it is
perhaps impoffible for human abilities
ever to difcover all their virtues and
properties.   A compofition of various
ingredients often effects what all of
them fingly would attempt in vain :
and as thefe combinations, as well
as the fimples themfelves, are infinite,
we need not be furprifed, that notwithftanding the prodigious number
of eminent men, from the age of
Hippocrates to our own times, have
fpent their lives in improving the healing

ing art, it is ftill far from having reached the fummit of perfection : and that we are yet ignorant of the virtues of many fimples with which the bountiful hand of nature has fo beautifully adorned the furface of the earth. New medicines are every day difcovered, and chance or accident often reveals what the moft affiduous application might have fought in vain.

Perhaps a ftronger inftance cannot be given of our limited knowledge of medicines, than the many ineffectual attempts to fubdue the inveterate malignancy of fome diftempers, which have hitherto withftood all the power of phyfic, and are ftill confidered as the opprobria of the healing art. But this muft not be imputed to nature ; fhe has provided medicines fufficient for every diforder: difeafes of this kind

A 2   triumph

triumph only over the ignorance of the practitioner.

Among thefe maladies, venereal and fcorbutic complaints, efpecially when complicated, are not the leaft formidable ; and phyficians have often confeffed, and often lamented, that they have too frequently baffled the power of every known medicine, and defied every attempt to fubdue their violence. If my own declaration will add any weight to the general voice of practitioners, I can, with great truth, affirm, that during a long feries of affiduous practice, I have met with numberlefs inftances, where all the common methods have been tried without fuccefs, and the wretched patients perifhed under the excruciating tortures of a loathfome difeafe. The fact indeed is too evident to want any farther proof: the univerfal acknow-
ledgment

ledgment of the moſt celebrated phyſicians have placed it beyond a doubt.
Should any perſon, however, ſtill heſi-
tate to ſubſcribe to the general opinion,
I could wiſh him to viſit the places
where patients of this kind are received;
becauſe ocular demonſtration would
there convince him, that the faſt is too
certain. He would there ſee objeſts labouring under the extremity of anguiſh,
breathing the moſt putrid exhalations,
and quivering on the brink of eternity,
after every known method has been tried in vain.

Affeſted by ſufferings which could
not be mitigated, and pierced with the
cries of diſtreſs which could not be relieved, I have often wiſhed that ſome
fortunate accident, or ſome happy ge-
nius, might diſcover a remedy, which
would enable the faculty to conquer
theſe

thefe dreadful difeafes. But the wifh, however generous, was long in vain; and I began to defpair of ever feeing it accomplifhed, when I fortunately became acquainted with M. de Velnos, the author of a medicine which has acquired great reputation at Paris, as a fpecific in venereal and fcorbutic diforders, either fingly or complicated. I had been too often deceived by pretended difcoveries, to believe implicitly his moft folemn affeverations, with regard to the great virtues of his medicine. I had never been an advocate for fecret remedies, and feared that intereft, rather than truth, might form the bafis of his affertions; and that though it might poffibly fucceed in flight cafes, it might want power to eradicate the difeafe when grown too powerful for other remedies. He perceived my doubts, and took an effectual method to remove them:

them : he offered to fubmit the efficacy and merit of his medicine to a feries of fair and candid trials. This was a propofal which no ingenuous practitioner could refufe. I accepted the offer ; and, from a fufficient number of cures performed on patients labouring under thofe dreadful diforders, fome of them confidered as abfolutely incurable, I am convinced, that his affertions were founded on truth ; and have the ftrongeft reafons to hope, that his medicine will prove of the utmoft advantage to this country. For whatever oppofition it may meet with, either from intereft, or bigotry to the old methods of practice, I am perfuaded its own merit, when fairly tried by the unerring touchftone of experience, will rife fuperior to all oppofition, and filence even the tongue of envy. The candid and ingenuous, thofe who can feel for the fuffer-

ings

ings of their fellow creatures, and fym-
pathize with diftrefs they wifh to relieve,
will, I am fatisfied, give the medicine
a fair and impartial trial.

I am not ignorant that every perfon
who attempts to recommend a medicine to the public, without revealing
the ingredients of which it is compofed,
muft expect the moft illiberal attacks
from the malignant pen of calumny.
His conduct will be feverely cenfured
by thofe whofe intereft is concerned to
fupprefs the difcovery ; and perhaps he
may not efcape the animadverfions of
others, though they may have only
fome favourite theory, or even the old
method of practice, to defend. But I
have learned to defpife all fuch ungenerous attempts : they are the common
attendants of any innovation. Almoft
every member of the faculty, however
eminent

eminent for his parts and learning, has met with illiberal treatment on endeavouring to introduce any new difcovery, new medicine, or new operation in furgery. It will be needlefs to mention examples of this kind : he muft be ignorant of medical hiftory indeed, who does not immediately recollect many that have happened both in our own and foreign countries. It fhould alfo be remembered, that feveral of the moft eminent phyficians, both of the laft and prefent centuries, had their fecret remedies ; and perhaps they did more fervice to the community by concealing the compofitions, than they would have done, had they publifhed them to the world.

It is well known, that M. de Velnos performed, by the help of his medicine, amazing cures during his fhort

ftay

ftay in England: I myfelf was a wit-,
nefs to feveral. I faw, with a fecret
fatisfaction, his fuccefs, and became an
advocate for his remedy from convic-
tion. I have alfo received inconteftible
evidences of its prodigious efficacy
from feveral of the firft nobility in
France ; together with certificates from
fome of the moft eminent phyficians
in that kingdom. Thefe, I hope, will
be confidered as proofs fufficient to ex-
culpate me from entertaining even the
moft diftant wifh to impofe upon the
public, in a matter of fo much importance. No perfon furely, who, by the
affiduous labour of many years, has acquired fome fhare of reputation, and
eftablifhed his character as a man of
veracity, will rafhly engage in an un-
dertaking, that may fap the founda-
tion of a ftructure, which has coft him
his whole life to erect! For whoever
publifhes

publifhes a treatife of any kind, fub-
mits the truth of what he has advanced
to the examination of the public ; and,
confequently, cannot complain of its
being thoroughly fcrutinized. If his
affertions are founded on truth, he has
nothing to fear ; like gold from the
furnace, they will acquire a greater luf-
tre from a candid enquiry.

THE

THE

INEFFICACY

OF ALL

Mercurial Preparations, &c.

———————

THOUGH Commerce is undoubtedly an inexhaustible source of wealth, yet population forms the genuine riches of a kingdom; it gives at once both strength and permanence to the state. Without the assistance of a sufficient number of people, the trade of any country must languish and decay: the boasted manufactures must be neglected, and even the necessary labours of the husbandman must fail. But notwith-

ſtanding the evident ſuperiority of population, it has been much leſs regarded than
commerce; which, in this land of liberty,
has been carried to a degree of perfection
unknown in other countries, by the in-
duſtry, diligence, and activity of the inhabitants, protected by the power, and aſ-
ſiſted by the countenance, of the govern-
ment. Every manufacture has been improved, and every invention encouraged,
that had a tendency to promote the various
branches of trade carried on in this flouriſhing
kingdom. Theſe are doubtleſs noble efforts,
and demand the applauſe of every lover of
his country: but ſurely population, on
which even commerce itſelf depends, ſhould
not be neglected. It is really aſtoniſhing,
that in a country, where the advantages and
diſadvantages of every contingent are nicely
calculated, and the effect of almoſt every
cauſe foreſeen: where the relief of human
nature, and the general advantage of ſociety,
form the ſole wiſh and aim of the legiſlature;
of

fo little regard is paid to the prefervation of the lives of men, though confefledly one of the grand fources of population. Luxury, and difeafes of various kinds, combine their malignant influences in leffening the number of the human fpecies. The exceffive ufe of fpirituous liquors, and other modes of living, are deftructive of propogation. The continual emigration of people to America, the vaft number of hands employed in fhipping, and the late call for men to the Eaft Indies, and to fettle our new conquefts, tend greatly to leffen the populoufnefs of this country, to the irreparable injury of agriculture and commerce.

But among all thofe caufes of depopulation, that relative to difeafes is the moft alarming; becaufe its deftructive effects, though amazingly rapid, are filent and fecret. How many daily perifh by venereal and fcorbutic difeafes: a lofs of the utmoft confequence, becaufe it greatly falls upon

thofe

thofe men who are of the greateft fervice to the ftate. The fcurvy may be confidered as epidemical to England, owing perhaps to its fituation as an ifland, the coldnefs and moifture of the air, the food of the inhabitants confifting chiefly of meat, and the long voyages at fea, performed by fuch numbers of people. We need not therefore be furprized to find, that almoft every native of this ifland is more or lefs afflicted with the fcurvy. This difeafe is of itfelf a fufficient misfortune, but when complicated with a venereal complaint, becomes truly alarming. The latter indeed is not, like the fcurvy, natural to this country; it has been imported from the continent to the deftruction of thoufands; and may perhaps be confidered as one of the evils refulting from a foreign trade. Wealth, the child of commerce, is the parent of luxury and pleafure; and hence the venereal difeafe, too often complicated with the fcurvy, acquires a malignity in this ifland, which renders it more violent, rapid;

and

and fatal. Thefe difeafes, handed down to pofterity, are poffibly the caufe of that degradation of the human fpecies, which had been fo often lamented by the writers of our own times. Every difcovery, therefore, which tends to leffen this deftructive malignancy, certainly deferves the attention of the public. It is, perhaps, abfolutely impoffible to deftroy entirely 'the fource of this evil; but it is poffible to ftop its alarming progrefs, if a medicine can be difcovered that will eradicate both venereal and fcorbutic diforders, whether fingle or complicated. A difcovery of the moft fingular importance to our foldiers and feamen, afflicted with the former, becaufe it is almoft conftantly aggravated by a natural fcorbutic habit.

It is well known, that the venereal difeafe is more eafily cured, both in France Italy, than in England; an advantage that refults entirely from the warmth of the cli-

mate;

mate; yet even there the difease too often becomes incurable, and triumphs over all the art of the phyfician. It is therefore no wonder that the difficulty is doubly increafed in this country, and that the faculty have long wifhed that a medicine could be difcovered, which would enable them to conquer this dreadful difease in its moft malignant ftate, even when complicated with the fcurvy. And this noble difcovery has, fortunately for the public, been made by Mr. de Velnos. His Antivenereal Vegetable Syrup, radically cures every ftage of the venereal difease, though complicated with a violent fcurvy.

But, before I proceed to enumerate the virtues and properties of this valuable medicine, it will be neceffary to confider the methods generally made ufe of in the cure of fcorbutic and venereal difeafes, and fhew wherein they are deficient and inadequate to anfwer the intended purpofe.

It

It is not sufficient, in the cure of the scurvy, to preserve the habit of body from the predifposing caufes: a long methodical treatment is neceffary.

The curative indications confift in correct-ing or preventing the alcalization of the humours; reftoring the tone of the folids, and the fluidity of the juices; re-eftablifhing the fecretions; and, laftly, in procuring the neceffary evacuations.

The medicines commonly made ufe of to effect thefe purpofes are fweet, acid, and bitter vegetables, given in different forms, and in different quantities, according to the various ftages of the diforder. But thefe are inadequate to the purpofe. It is not fufficient to affect an alteration in the humours, or even to re-eftablifh the natural functions; for, unlefs nature be farther affifted by proper evacuations, either by ftool, fweat, or urine, the difeafe too often degenerates into a

dif-

difeafe of another fpecies; when it will be neceffary to have recourfe to antifcorbutics, given as alteratives, purgatives, bitters, acids, corroboratives, and gentle fudorifics.

Thefe medicines, which experience has fufficiently proved to be neceffary in making a radical cure, have their inconveniencies. Purgatives, however properly chofen, and artfully managed, too often weaken and impede digeftion; and even their operation is not fupported without difficulty by perfons of fcorbutic habits. Sudorifics difturb the operations of nature, and, finding the humours in a ftate approaching to a difiolution, increafe the evil, and often occafion a total decompofition. Hence we fee the reafon why the common method of curing the fcurvy, though remarkably tedious, is not free from danger; nor is it always certain: too many inftances occur in medical hiftory, where all the abilities of the phyfician have been exerted in vain.

The

The venereal difeafe is too well known, and its effects too often fatally experienced, to need any explanation, either with regard to its nature, its origin, or the manner of. its communication. It will be fufficient for my purpofe to confider the feveral medicines hitherto made ufe of in curing this difeafe, and fhew that they.are all of them too often inadequate to effect the neceffary purpofe.

Three different kinds of remedies; fudorifics, purgatives, and mercury, fometimes alone, and fometimes compounded, have been hitherto ufed in the cure of this diforder. Experience has abundantly fhewn that the two firft are not fufficient to anfwer the intention. Mercury, notwithftanding all its inconveniencies, has indeed long been confidered as a fpecific in the venereal difeafe; but in my opinion very unjuftly. I wifh there were no inftance upon record where all its boafted virtues have failed, the patients, after taking it in almoft every form,

have

have fallen a facrifice to this loathfome difeafe. And perhaps the only reafon why it
has fo long maintained its place in practice is,
becaufe no other medicine of equal virtue,
in venereal complaints, has hitherto been
known.

It will not be denied but mercury often
cures the venereal difeafe; but is not the
patient expofed to dangers and accidents of
the moft alarming kind ; and is not the cure
too often partial and incomplete ?

Mercury is adminiftered externally and
internally.

The external methods are by friction and
fumigation. By either of thefe, it is divided
into very minute particles, penetrates through
the pores in its metallic form, and affects the
falivary glands.

By this method we are never certain with
regard

regard to the quantity; we cannot judge of it by its effects, becaufe thefe are equivocal; it never acts equally on different patients; and the diffolution of the fymptoms is an uncertain fign of a radical cure; for it fometimes happens, that notwithftanding all the fymptoms occur, the caufe is not totally removed. What dependence therefore can be placed upon a medicine, whofe effects are fo uncertain? If we add to this the danger to which the patient is expofed, notwithftanding all the care of an experienced and prudent phyfician, our opinion of the virtues of this boafted remedy will be greatly leffened. How often do the moft fatal confequences happen from its ufe? It is not uncommon for the mercurial particles to be diverted fuddenly to the head, ftomach, or inteftines, where the ravages they make are abfolutely irreparable.

Whether the intention be to falivate, or whether the falivation comes on, notwithftanding

ftanding all the care and precaution of the phyfician, the following fymptoms generally attend it : a fwelling of the head and neck ; an obftruction and irritation of all the falivary glands : pain, inflammation, and exulceration of the internal parts of the mouth, attended wtih flough or excrefcences, which occafion hemorrhages more or lefs confiderable ; fœtid gums; loofenefs and lofs of teeth ; and a cruel want of fleep, occafioned by the fear and danger of fuffocation.

The patient is more or lefs expofed to thefe dreadful fymptoms, as he is more or lefs afflicted with a fcorbutic habit ; while the natural effects of the falivation, a ftrict regimen, pain, and an almoft total want of fleep, reduce him to the loweft ftate of depreffion ; without affording him the confolation, that a radical cure will be the confequence of his fufferings : becaufe mercury, though adminiftered with the utmoft caution to patients afflicted with the fcurvy, brings on too fud-

denly

denly a falivation, which is very difficult to
ftop; and the quantity of mercury the patient has received is not fufficient to deftroy
the venereal virus. Nor can this defect be
fupplied; for every time an attempt is made
to increafe the quantity, the fame accidents
return. When this happens, and the venereal
fymptoms are alarming, the cafe of the patient
becomes deplorable; the fcurvy is irritated
by a medicine abfolutely improper, and both
difeafes foon increafe to a degree that almoft
defies the power of medicine, and leaves the
miferable patient without refource.

But the fcurvy is not the only circumftance
that renders a falivation improper in the cure
of the lues venerea. There are others, among
which we may reckon various diforders inci-
dent to the palate, the uvula, and the tonfils; together with the erofions and exulce-
rations of the fauces, which, fpreading deep
into the fubjacent parts, often refemble an
eating cancer, which cannot be totally cured

and

and confolidated, till the diforder which gave them birth is effectually fubdued and removed. In this cafe, therefore, if in any, we are to obferve the following maxim, that the peccant matter is neither to be conveyed to, nor evacuated by, the part affected; other-wife it muft neceffarily happen, that by an accumulation of the faliva, which to the tafte of the patient is æruginous, virulent, and has its qualities heightened by the mercury, a gangrene muft be produced, and foon after fucceeded by a fatal fphacelus of the parts. Hildanus furnifhes us with fome remarkable inftances of this kind, *Cent.* 3. *Obf.* 92. Salivation is alfo improper in cafes where, befides a redundance of thick and vifcid hu-mours, the patient's ftrength is much impair-ed; and this is certainly a very juft and rati-onal maxim; fince, in confequence of the langour of the whole body, and the defect of a due tone, and fufficient motive force, all the parts, and even the more noble vifcera are preternaturally flaccid; hence it happens, that

that the vifcid fordes of the humours being thrown into violent commotions by the efficacy of the mercury, are eafily, and in great abundance, thrown into thofe vifcera; but cannot, in confequence of the weaknefs of the refifting fibres, be fo eafily expelled from them. Hence thefe accumulated and peccant humours become ftagnant, and occafion terrible fymptoms of various kinds.* But a more irreparable misfortune is produced, when thefe fordes are conveyed to the brain, already weakened by previous diforders; for by this means, palfies, apoplexies, and other terrible lethargic diforders muft be foon produced. But fince in a lues venerea of fome ftanding, and deeply rooted in the humours, there is generally a large quantity of peccant matter; and fince the ftrength of the patient is, for the moft part, much exhaufted, either by the force of the difeafe, previous intemper-

---

* Vide Sennert. Prax. Med. Part. 4. Lib. 6. Fallop. De Lue Venerea: & Sylv. Math. Med. Lib. 2. Cap xi.

ance,

ance, furfeits, or an excefs of venery, it generally happens, that, in thofe circumftances, a falivation, by fome confidered as the only fovereign remedy for a lues venerea, is highly improper and abfurd. Many are indeed of opinion, that even in thefe circumftances the body may be rendered fit for bearing a falivation; if, for inftance, before its ufe, the redundance of the peccant matter is leffened, and the fpiffitude corrected by venefections, purgatives, and the repeated exhibition of fudorifics; but Sydenham's opinion is certainly much better founded; that ingenious practitioner, in his Treatife *De Lue Venerea*, tells us, that by fuch meafures the body is no more prepared for bearing a falivation, than the bodies of foldiers would be prepared for battle, by cutting their nerves. Hence we fee that mercury ufed externally, either by rubbing or fumigation, is fo far from being a fpecific in the cure of venereal diforders, that it is often deleterious; inftead of curing, it deftroys the patient. Let us now confider

confider how juftly it may be termed a fpecific in the cure of the lues venera, when admi‑ niftered internally.

Mercury in its crude ftate, and undivided by any intermediate body, given internally, produces no fenfible effects. Cinnabar and Æthiop's mineral, very little; becaufe it is abfolutely neceffary to reduce mercury to a faline form, in order to render it capable of being diffolved in the juices of the ftomach; for by that means only it is capable of enter‑ ing the minute veffels of the human body. On this principle, mercury has been com‑ bined with all the mineral and vegetable acids, and thence an infinite number of fa‑ line mercuries have been produced; but the effects and dangers of them all are directly proportional to the quantity of acid fpiculæ, with which the mercury is loaded. Hence we fee the reafon why mercurial preparations degenerate into poifons; and why they are always more or lefs dangerous.

The

The principal preparations of mercury, given in a dry form, are turpeth mineral, mercurius dulcis, calomel, panacea of mercury, white precipitate, &c.

Keyfer's and Bellofte's pills, by their difficult folubility, remain long enough in the ftomach to occafion the moft alarming mifchiefs; and thefe will always be proportional to the degree of activity of the medicine, and the quantity of the dofe. This method of exhibiting the faline preparations of mercury, fhould never be practifed without the utmoft caution, on account of the alarming accidents that may refult from their extreme irritating properties.

Sometimes, on account of the violent corrofive quality of the mercurial preparation, it is difolved in a large quantity of fome fluid, as water, fpirit of wine, &c. Even corrofive fublimate, one of the ftrongeft preparations of mercury, is given in this manner; and fold

to

to the public under different names; but always to the prejudice, and too often to the
total deftruction of the patients conftitution.
By this method of exhibiting the faline preparations of mercury, its particles, being already minutely divided, penetrates the ca-
pillary veffels, and at once irritates and cor-
rodes the nerves. It is therefore no wonder
that the afflicted patient, too often with a painful degree of certainty, attributes to this caufe
the firft attack of a nervous diforder, which
all the power of phyfic can never remove.

If to thefe we add the inconveniences infeparable from mercurial preparations, and
which are always the confequences of their
ufe; if we confider the fmall number of patients to whom mercury may be exhibited
with fafety, among whom we muft rank
youths and old men, women and hypochon-
driacs; becaufe the delicacy and fenfibility of
the nerves of the one, and the rigidity of thofe
of the other, render them more fufceptible of

the

the bad effects of saline mercury; the use of mercurial medicines will be confined within very narrow limits. It frequently causes miscarriage in women during the firſt months of their pregnancy, and kills the fœtus in the womb of thoſe who are farther advanced; it debilitates the ſtomach, occaſions dyſenteries and ſpitting of blood; cauſes ſevere and continual pains, tremblings, and the palſy: accelerates a pulmonary conſumption in thoſe a little inclinable to that diſorder: for mercury always greatly affects the lungs of thoſe who have made a free uſe of it. Add to this the chronicle diſeaſes occaſioned by this medicine, and which are with the utmoſt difficulty, if ever, cured; and generally ſuppoſed to owe their origin to other cauſes; its being abſolutely improper to be adminſtered to valetudinarians, and perſons of weak and delicate conſtitutions; and its being abſolutely contrary when the diſeaſe is complicated with others, eſpecially with the ſcurvy; and then let the reader judge how far mercurial preparations are uſeful. In

In a variety of venereal cafes, mercury, inftead of removing, increafes the complaint;
fuch as ulcers in the palate, fwelling of the
tonfils, caries and exofto'es of the bones, in-
durations of the glands of the groin and
neck, buboes, where they have acquired a
certain degree of malignancy, inflammation
and hardnefs of the tefticles, warts, and frefh
contracted claps; in many other cafes it
ferves only as palliative.

Where then is the boafted efficacy of mercury? Experience has abundantly proved that
the fmall degree of advantage derived from
mercurial preparations is more than balanced
by the alarming confequences infeparable from
their ufe: the inconveniences far exceed the
utility.

But it may be afked, how can thefe confequences be prevented? Is it poffible to cure
fome ftages of the venereal difeafe without
the affiftance of mercury? The anfwer is
eafy: It is---Velnos's Vegetable Syrup effec-

tually

tually performs the cure, without being followed with any of thofe evils, which always attend a courfe of mercurial medicines. It totally eradicates the difeafe in its moft alarming ftages, without offering the leaft injury to the conftitution. It may be given with the utmoft fafety to perfons of the moft delicate frame; to youths, and even women far advanced in their pregnancy, Its action is remarkably gentle: it ruffles not the conftitution; it impedes not the common operations of nature. Of benign and friendly properties, it rather fupports than impairs the nervous fyftem. It is equally effectual in the fcurvy as in the venereal difeafe; a complaint which has fo long baffled the fkill of the moft eminent phyficians, and which is fo common in this country. It anfwers all the curative indications already enumerated for eradicatin t h at cruel difeafe; it ferves at once as a remedy for the diforder and food for the patient. It acts as an alterative, and by its corroborative qualities re-eftablifhes the fecretions; difcharges part

of

of the peccant matter through the pores of the ſkin; looſens the belly to a convenient degree, and becomes a purgative without its inconveniences.

But when the venereal diſeaſe is complicated with the ſcurvy, the Vegetable Syrup of M. de Velnos, is perhaps the only remedy hitherto known that can effect the cure. It is equally adapted to both, and both are eradicated with the ſame eaſe, and nearly in the ſame time.

A medicine, endowed with theſe noble virtues, ſurely deſerves the notice of phyſicians; as it will enable them to conquer a complication of diſeaſes, which has hitherto defied the power of medicine, and carried thouſands of uſeful members of ſociety untimely to the grave. This capital medicine has been known ſome time in France, and has met with the approbation of the faculty at the royal college of phyſicians in Paris, and I would

willingly

willingly hope, that by introducing here a medicine, which totally eradicates all kinds of venereal and fcorbutic diforders, without the leaft injury to the conftitution, I fhall be thought to have done fome fervice to this country, by faving the lives of many of my fellow creatures.

But in order to fupport in England the great reputation which the Vegetable Syrup of Mr. de Velnos has juftly acquired in France, and prevent the public being im-pofed upon by a fpurious medicine, it is ne-ceffary to obferve, that a compofition of a very different nature is fold in London, under the deceptive title of *The Vegetable Syrup of Mr. de Velnos, with improvements.* Dr. Burrows, the author of this medicine, being well acquainted with the virtues and efficacy of Velnos's Syrup, and the great reputation it had juftly acquired at Paris, agreed with the inventor for a certain number of bottles of the Vegetable Syrup, which he adminiftered

in

in London, promifing to pay M. de Velnos a certain price for each bottle, and actually performed fome remarkable cures with that quantity of the medicine. But his covetoufnefs foon proved too ftrong for his virtue; he fcrupled not to facrifice the confidence of a generous public at the fhrine of avarice. A fpurious compofition, under the ungenerous pretence of its being the medicine of Mr. de Velnos with improvements, has for fome time been advertifed in the public papers by this *confcientious* gentleman.. Dr. Burrows has indeed pretended, that he knew the compofition of Mr. de Velnos's Vegetable Syrup. But it may furely be afked how he came by the fecret? he never had it from the inventor; and no other perfon in the world knew what the ingredients are of which it is compofed; before Dr. Mercier and myfelf, who purchafed the receipt from Mr. de Velnos.

It

It has more than once been afferted in this differtation, that Mr. de Velnos's Syrup is compofed of fimple vegetables only; and that from this circumftance it derives a peculiar advantage over all other medicines extracted from any bodies belonging to the mineral kingdom. Becaufe, though it is far more efficacious, it is not fufceptible of thofe dreadful confequences that generally attend a courfe of draftic mineral preparations. It may be given with fuccefs in all complications, at all feafons, and to all conftitutions. Even an error in the adminiftration, or the imprudence of the patient, can produce no alarming effect. It is therefore of the utmoft importance to eftablifh the truth of this affertion, namely, that the Syrup of Mr. de Velnos is compounded of fimple vegetables

only:

only : and I flatter myſelf that the following A N A L Y S I S of the Medicine
will be abundantly ſufficient for that pur-
poſe.

ANALYSIS

A N A L Y S I S

Of the Syrup of Mr. de Velnos, * *made by Order of the* Marſhal Duke de Biron, *by* Meſſrs. Rouelle *and* La Caſſaigne, Pro-feſſors of Chemiſtry in Paris; *who firſt made the Experiments ſeparately, and then, in order to corroborate the Truth of the whole, repeated them together.*

THIS Remedy, in its natural ſtate, exhibits the appearance of a ſyrup of a thinner confiſtence than the common; of a brown colour; tranſparent; of a taſte a little

---

* This Medicine was adminiſtered by Dr. Mittie, in the hoſ-pital of French guards, under the infpection of Meſſrs. Bercher, Le Thuilier, Doyen, Guilbert, and Deſlou, doctors regents of the faculty of medicine, and M. Dufouart, furgeon major of the regiment of guards.

little medicinal, but principally that of fugar.

Experiment I.

ONE pound of the Syrup, diftilled in balneum mariæ, gave not the leaft indication of fpirit; but the phlegm was flightly aromatic.

Experiment II.

FOUR ounces of the Syrup poured into a plate of delph ware, and placed in balneum mariæ, produced one ounce five drams of cryftaline matter, of a brownifh colour, and which attracted the humidity of the air.

Experiment III.

SIX drams and a half of the above cryftalline matter, being put into two ounces of

F   rectified

rectified spirit of wine, and digested in balneum mariæ, the menstruum was strongly
tinctured; and, after being decanted, filtrated
and set in a cool place, deposited a sweet,
viscous extract.

Experiment IV.

A second digestion in the same quantity of
spirit of wine, gave the same produce.

Experiment V.

BY a third digestion in the same quantity of a spirit of the same quality, the menstruum was not so deeply tinctured; and,
being deposited in a cool place, a small
quantity of little crystals of a sweet sugary
taste, shot to the sides of the vessel.

Experiment VI.

A fourth and fifth digeftion gave the fame produce.

Experiment VII.

AFTER thefe five digeftions, there remained a black matter weighing twelve grains, of an infipid tafte, abfolutely infoluble in fpirit of wine, but eafily foluble in water. On being thrown on burning coals, it gave a fmell compounded of thofe arifing from animal and vegetable fubftances; owing to the matter ufed in the clarification of the Syrup.

Experiment VIII.

THE menftruum charged both with the extract and faline matter, being mixed, firft
with

with a little, and afterwards with a large
quantity of water, did not become turbid.

Experiment IX.

THE spirit of wine, charged with the
extracted matter of the first and second digestion, being drawn off in balneum mariæ,
and again cohobated and distilled, there remained an extract, sweet and very tenaci-
ous, of a yellow brown colour, attracting the
moisture of the air; and being added to that
which was actually deposited, weighed three
drams and a half.

Experiment X.

WE have already observed, that from
the third, fourth, and fifth digestions, some
crystals were obtained; these crystals weighed a dram. The menstruum in which they
shot,

ſhot, being evaporated, produced no more; but there remained a dry ſugary matter, weighing one dram and one ſcruple. The cryſtalline particles were prevented from ſhooting, by being wrapped up in a ſmall portion of the unctuous extract obtained by the two firſt digeſtions.

Experiment. XI.

THESE cryſtals, as well as the unctuous extract, being thrown upon burning coals, did not decrepitate, but diffuſed a ſmell of burnt ſugar.

Experiment XII,

FOUR ounces of the Syrup, being placed in balneum mariæ, and the fluid evaporated to a dryneſs, there remained one ounce five drams of deſiccated matter. This matter pul-

pulverifed, was put into a glafs retort with four ounces of well rectified æther, and digefted in a fand heat, with a moderate degree of fire. The æther was then drawn over into a mattrafs luted to the mouth of the retort; cohobated; again diftilled, and again cohobated; after which the digeftion was continued for two hours. By this means we obtained a tincture, diffufing the fmell of ambergris; and the æther being evaporated, the dry matter had the fame fmell, was tranfparent, and of an acrid tafte. This deficcated matter diffolved totally in fpirit of wine; and water being added to the tincture, it rendered it but very little turbid.

Experiment XIII.

A part of the above deficcated matter digefted in warm water diffolved totally, except a few minute refinus particles in fuch fmall quantity, that they feemed not to have refided

fided originally in the medicine, but to have been formed by the action of the fmall portion of acid in the æther on the oily parts of the vegetables ufed in the compofition of the fyrup.

Experiment XIV.

SIX ounces and a half of the dificcated matter procured from a pound of the fyrup, being put into a retort, and the veffel placed in a reverberating furnace, there firft came over a confiderable quantity of phlegm, the acidity of which increafed in proportion as the fire was augmented. As the heat became more violent, the acid became more concentrated, and more empyreumatic; and at the fame time there came over a black oil, partly limpid, and partly thick. The fire being raifed to the moft violent degree, more acid and oil came over; but not a fingle particle of mercury. The retort being broke,

there

there remained, difperfed all over the internal fource, a burnt matter or coal, extremely rarified, like that of an unctuous
body.

Experiment XV.

THIS coaly matter being pulverized,
mixed with the black flux in a crucible
well luted, and melted in a wind furnace,
not a fingle-metallic particle was found at
the bottom of the crucible.

(Signed) ROUELLE.

LA CASSAIGNE.

(Profeffors of Chemiftry in Paris.)

O B-

Obfervations and Deductions.

I.

IT follows from the firft experiment that one or feveral aromatic vegetables are ufed in the compofition of the Syrup.

II.

From the fmall degree of colour and tafte of the extract obtained by the fecond experiment, it is plain, that the fyrup contains only a fmall quantity of active particles, and that fugar, or fome faccharine matter, forms the predominant part of the medicine.

III.

It follows alfo from the third, fourth, and ninth experiments, that the active par-

G   ticles

ticles in the fyrup are very few, and that a confiderable quantity of honey is ufed in the compofition.

IV.

From the fifth, fixth, tenth, eleventh, and fourteenth experiments, it follows, that fugar is alfo ufed in the compofition of the fyrup; becaufe we have been able to feparate it from the extract by cryftalliza- tion, and confequently to determine the quantity; and as it is very probable, that equal parts of fugar and honey are ufed, and as two drams and a fcruple of fugar are con- tained in two ounces of the fyrup, if we add two drams and a fcruple of honey, there will remain only one dram and twelve grains of active particles.

V.

It is evident from the feventh experiment, that the fyrup had been clarified either with ifinglafs,

ifinglafs or the whites of eggs ; but the quantity of either is too fmall for us to imagine it
is ufed to increafe the virtues of the medicine.

VI.

From the eighth, twelfth, and thirteenth
experiments it appears, that the fyrup does
not, in its natural ftate, contain any refinous
particles; but as the active part diffolves
equally in water and fpirit of wine, it may
be confidered as a refinous extract.

VII.

In fine, it is evident from the fourteenth
and fifteenth experiments, that the fyrup
does not contain a fingle particle, either of
mercury, antimony, or any metallic or femi-
metallic fubftance whatever.

CONCLUSION.

From the whole therefore it appears that
the fyrup of Velnos is compounded of a confiderable

fiderable portion of fugar and honey, of a refinous extract from one or feveral aromatic vegetables, and a fmall part of ifinglafs or whites of eggs; the whole diffolved in a large quantity of water.

Signed

LA CASSAIGNE,
AND
ROUELLE.

CASES

Dr. BURROWS's late PAMPHLET.

HOwever unwilling the author of the following little tract has always been to
anfwer declamatory language, fuch calumny
obliges him to prove, before an impartial public, his own rectitude and his opponent's bafe-
nefs. When falfehood and detraction are
employed to vilify an individual, a regard to
his own character admonifhes him to reply,
but when that falfehood may eventually in-
jure the community, it is criminal to be
filent.

The

The author, confcious of the juftice of his caufe, with pleafure appeals to facts, difdaining to ufe fcurrility for argument, or fallacies for truths. In a pamphlet publifhed by Dr. Burrows, with the pretence of refuting arguments advanced in a publication by the late Mr. Saffory, furgeon, the readers were promifed a full and fatisfactory account of the whole matter in difpute, and the title page, like the trumpeter at a puppet-fhew, announced that his was the only booth in the fair; but expectation, as ufual, was difappointed, and the piece refuted itfelf.

A difputed point is feldom difcuffed with that candour and calmnefs neceffary to inveftigate truth. Men are often deluded by their paffion, and florid futile arguments often deceive the underftanding, and miflead the mind. Falfe logic, like a glaring blaze, dazzles the fenfes, but is foon extinguifhed by the fteady beams of truth. It therefore is neceffary to reply to the many falfe affertions

of

of Dr. Burrows, detect his evasions, and ex-
pose him to the public, as he realy is, an *im-
postor. Fænum habet in cornu, longe fuge.* HOR.

Well knowing the difficulty of discounte-
nancing conceit, or reclaiming injustice, the
author wished to decline the thankless task ;
but as the *Doctor*'s self-sufficiency has brought
this retort upon himself, let him remember
that he who throws the first stone is answer-
able for the consequences.

Dr. Burrows informs us that Mr. de Velnos was bred to the sea, and commanded a
ship trading to the Levant, and sagaciously
concludes that therefore he was not the original inventor of the Vegetable Syrup. Have
not the most useful discoveries both in physic
and mechanics originated either from chance,
or the experiments of the curious ? Mr. de
Velnos fortunately learned the basis of his
medicine in his travels ; he has acknowledged
that the salutary roots, plants and flowers, the
com-

component parts of his fyrup, were pointed out to him in Nigritia, by a native of Calabria, and being an eye witnefs of its wonderful effects in the moft obftinate venereal and fcorbutic diforders, he was induced to make it into an agreeable fyrup, of which he claims the invention.

Dr. Burrows afferts in his affidavit that he not only knows the compofition of Mr. de Velnos's Vegetable Syrup, but has improved the fame. Would not a candid declaration of the manner, time, and place in which he procured this knowledge fatisfy the public, more than the mere *ipfe dixit* of an interefted man ? Meffrs. Saffory and Mercier have every teftimony in their poffeffion, to convince the public of their being the only proprietors of Velnos's original Vegetable Syrup, as it will appear by the following affidavit :

A F F I-

AFFIDAVIT of MRS. de VELNOS.

JAne de Velnos, late of Paris, but now residing in Dean Street, Soho, in the county of Middlesex, maketh oath, and saith,
That she is the lawful wife of John Joseph
Vergely de Velnos, the inventor and sole pro-
prietor of a medicine known by the name of
Velnos's Original Vegetable Syrup; and this
deponent saith, That her said husband having
been afflicted with a severe and continued illness, hath long since intrusted this deponent
with the secret for making the said medicine.
This deponent hath since that time constantly
sold and administered the said medicine by
her agents in Paris, and elsewhere in the
kingdom of France, with the greatest success: And this deponent saith that her hufband having been prevented by such illness
from coming to England, she hath lately, by
virtue of a general power and authority given
to her by her said husband, communicated to

H   Peter

Peter Mercier, of Frith Street, Soho, in the county of Middlefex, doctor of phyfic, all and every the ingredients made ufe of by this deponent in the compofition of the faid Syrup, and hath taken the faid Peter Mercier into partnerfhip with her faid hufband and herfelf, for the better adminiftering the faid medicine in his Britannic majefty's dominions; the faid Peter Mercier having, in conjunction with Meffrs. Saffory, furgeons in London, been concerned in the fale and adminiftration of the faid medicine for three years paft; and this deponent upon her oath faith, That the faid medicine hath not any thing of a mercurial, antimonial, or metallic nature whatfoever in its compofition, but is made and prepared from fimple vegetables only; and this deponent faith, That fhe hath been fo induced to communicate the faid fecret to the faid Peter Mercier, as well in confideration of the great truft and confidence this deponent and her hufband have in his integrity, as alfo, if poffible, to prevent the public from

being

being impofed upon by any medicine offered to them in the name of this deponent's huf-band, every fuch medicine being *fpurious*.

Thus the *Doctor* hoped to overturn the ftructure of reputation raifed by the Original Vegetable Syrup, and to mount upon the ruin of others ; though he prefumes to fell as a principal the medicine he once fold as an agent, may as eafy detection ever attend fimi-lar impofitions !

Meffrs. Saffory and Mercier procured the fecret by a legal purchafe, at which this agent took great umbrage, but has never told the public how he obtained his knowledge of the compofition, and has conftantly refufed to fubmit his medicine to analyzation, the only fair mode of proving it genuine or fpurious.

Dr. Bur-

Dr. Burrows next affures us that Mr. de Velnos has long fince divulged his fecret to a committee of twenty-four members of the faculty at Paris: this he fhould know is a matter of form in that country, upon granting what they call a privilege. Mr. Senac, firft phyfician to the king, being fatisfied by the accounts of Monfieur Petit and other eminent phyficians, of the great efficacy of Mr. de Velnos's medicine, after it had been analyzed by two profeffors of chymiftry, under the infpection of the Marfhal Duke of Biron, granted him, gratis, what is here ftyled a patent; but were his majefty's letters patent here as hard to obtain, the king's fignature would never have been proftituted as in the prefent inftance, nor would Dr. Burrows have had authority to give fraud a fanction.

Dr. Burrows alfo communicated his fecret to a committee, but did he divulge the component parts of his fpurious noftrum? he did not; nor was it required of him. I beg the reader

reader to judge whether the ableft chymifts could, from what he thought proper to reveal, gather fufficient information to enable them to make his fyrup.

COPY *of* Dr. BURROWS's *Depofition, taken upon* Oath.

" I take of the eccoprotic or milder purge-
" ing plants, with a proper quantity of fal
" tartar, and let them infufe fome time;
" after which, I make ufe of antivenereal
" and antifcorbutic plants, moderately bruif-
" ed before expreffion, adding a quantity of
" the juice of dendelion; then depuration is
" neceffary, for freeing them from all hete-
" rogeneous matter, afterwards let them
" ftand fome time in a moderate digefting
" heat; then decantation is required, after
"which defpumation, by adding white of eggs
" to the fluid to be clarified, and when boiled
" to a proper confiftence, filtration or perco-
lation

" lation follows, by paffing without preffure
" the fluid to be purified through proper
" ftrainers, at length it is made into a fyrup
" for ufe."

Rifum teneatis Amici.

How deftitute of underftanding muft that
man be, who can ftoop to contradict his own
affertions! this, through neceffity, Dr. Burrows has recourfe to; he has repeatedly acknowledged the receipt of one hundred quarts
of fyrup from Paris, at two different times;
yet in his laft pamphlet he affirms with his
natural effrontery, that he never adminiftered
a fingle bottle of Velnos's Syrup in this kingdom. The contrary is the fact. Dr. Burrows
received fifty bottles of Mr. de Velnos's fyrup
from Paris, which was paid for, agreeable to
articles; when that quantity was difpofed of
he fent for fifty bottles more, for which he
never made any remittance, and this broke
off the correfpondence between him and Mr.
de Velnos.

When

When it becomes neceffary for a man to rail, it is injudicious to ftigmatize his adverfary with the very epithet which he himfelf deferves. From this, however, in the warmth of his defamation the Doctor fhrinks not, and brands another with the name of noftrum-monger general, when he was contractor at one time for three different noftrums from Paris, viz. with Mr. Valabreque for his Perfian drops, with Mr. de Velnos for his Vegetable Syrup, and with Mr. Gamet for his cure of cancers : the laft he advertifed here lately, fince which the Doctor has fortunately hit upon a new method of convincing himfelf and friends, of the credulity of the public, by improving the art of midwifery, and facilitating births.

The remarks which have fallen from the pen of Burrows relative to Dr. Mercier's having kept an academy, his decoying Mr. de Velnos from France, his acting behind the curtain, and perfonating the late Mr. Saffory, are little paltry infinuations, more worthy the
mouth

mouth of a valet, than the great improver of medicine. Dr. Mercier has had an university education, and has received his diploma, which conſtitutes him maſter of arts, and doctor of phyſic; he was a partner from the beginning, and known as ſuch; never attempted to perſonate another, but by a candid behaviour has gained the eſteem of his numerous acquaintance, and the character of an honeſt man.

Mr. de Velnos came to England unaſked to claim his right, and to expoſe Dr. Burrows for impoſing on the public a ſpurious medicine in the name of the author.

Dr. Burrows ſays in his pamphlet, page 15, "Mr. de Velnos finding himſelf totally "diſappointed in his mighty expectations, "and probably having diſcovered that he "was allured over as a mere tool to carry "on the machinations of others; as the "wiſeſt ſtep he had taken during that whole "tranſ-

" tranſaction, returned to France, promiſing
" to reviſit London, but has not made his
" appearance ſince."

The following plain tale will expoſe the
Doctor's talent of falſifying. In order to put
a ſtop to the practices of an impoſtor, Mr. de
Velnos repaired to London, where he met with
the reception he deſerved both from his friends
and the public; and, during a reſidence of three
months, properly exhibited the character of
Dr. Burrows, his unfaithful agent. He then
returned well ſatisfied to Paris, with an intention ſoon to reviſit England, but age and
infirmities prevented him from putting this
deſign into execution. He therefore ſent his
wife to London, with a ſpecial power of at-
torney, and ſhe has confirmed upon oath the
truth of the above, * and other particulars,
which would be ſufficient to put any man
but Dr. Burrows to ſhame and confuſion.

* Vide the affidavit, page 57.

I   I know

I know not whether the Dr. poffefles the fecond fight, or by what other extraordinary quality he became fo well acquainted with our agreement with Mr. de Velnos. How can he tell what quantity of Syrup was left with Meffrs. Saffory and Mercier? they can prove they had a large ftock when Mr. de Velnos returned to France, and have fince received two invoices at fuch a rate as has enabled them to fell it at one guinea per quart, which may ferve as an anfwer to his filly obfervations concerning the reduced price.

Dr. Burrows afks, what fatisfaction can the public receive from the analyzation of the medicine? they are affured that it contains not a grain of * mercury or other metallic fubftance. Dr. Burrows fays the fame of his boafted fpecific, but has forgot to prove it;

* A quality which can with truth and juftice be attributed to no other remedy in the fame difeafe. *Dr. Burrows's firft pamphlet, page five.*

a fima fimple affertion from a man whofe affidavit has been proved untrue will be diftrufted.

Dr. Burrows, in his account of cures, infertted the cafe of Mrs. Mary Boardman of the Cloifters, four months after her death; this cure was certainly complete, and we doubt not but his medicine has given the fame effectual relief to many others.

Dr. Burrows accufes Dr. Mercier of vending a fpurious fort of bougies, thereby invading the property of Mr. Lallier, Surgeon. Now, the faid Lallier is no furgeon, but was a fervant to Mr. Daran in Paris, where he learned to make bougies. Dr. Mercier has been intimately acquainted with Mr. Daran both in Paris and in London, and purchafed his compofitions from a perfon who lived with the faid Daran, who made his medicated bougies for a long time, as may be attefted on oath, if required.

How-

However, the accufation, had it been true,
was impertinent, and ftrongly marks the
weaknefs of a man cafting about for arguments of recrimination.

The author of this little tract has now
gone through Dr. Burrows's pretended refu-
tation of the late Mr. Saffory's remarks, &c.
has fhewn it to be a fhamelefs and evafive re-
cantation of former principles. He hopes
that if juftice is preferred to a fpecious fhew
of honefty and truth, the remedy and arguments will alike bear analyzation, while
Dr. Burrows ftands convicted as *an impoftor*
at the bar of that impartial public, whofe
favour the author wifhes to enjoy, as he may
deferve.

CASES,

C A S E S,

Wherein Velnos's Vegetable Syrup *has been administered with the most surprizing Success in venereal and scorbutic Cases.*

C A S E I,

London,
(to wit) } ELeanor Smith, now of the parish of St. Botolph, Bishopsgate ; maketh oath and faith, that she, this deponent, for some months past has been afflicted with the most dreadful symptoms of the venereal disease, attended with shankers and sores of the foulest kind, together with the most excruciating and constant pains in her head and limbs, which reduced her to so miserable a state of lowness, that she could not walk across the room without assistance. This deponent also declares, that she was, at the same time, afflicted with what is called a hectic fever, in consequence of the above complaints,

complaints; but by the taking of M. de Velnos's Vegetable Syrup, adminiſtered to her by Meſſrs. Saffory and Son, ſurgeons in Biſhopſgate Sreet, London, this deponent declares, that ſhe was perfectly and radically cured ; that ſhe now remains in perfect health, and that all her ſores are healed, without having any other remedy applied to them, except the ſyrup uſed as a waſh.

ELEANOR SMITH.

Sworn, *September* 30, 1772.
> Before me,
WILLIAM NASH, Mayor.

C A S E II.

Mary Wilſon, late of the pariſh of St. Botolph, Biſhopſgate, about 22 years of age, laboured under a complication of ſome of the moſt obſtinate venereal ſymptoms I ever ſaw. The injury had been received

ceived fome years; and fhe had been long afflicted with the moft excruciating pains, running fores in many parts of her body, fhankers upon the labiæ, and a node of the fize of a pigeon's egg, and exceffively painful, on one of her legs. She was naturally of a fcorbutic habit, and this, complicated with the venereal difeafe, had fo emaciated her body, that even an attempt to effect the cure, by the common method, muft have been attended with the moft fatal confequences. But by taking of four bottles of Velnos's Vegetable Syrup, and obferving a ftrict regimen, every fymptom of her complaint was removed; fhe was perfectly cured, and in lefs than three weeks after recovered her former ftrength.

Witnefs to the Cure,
Le Febvre, M. D.
Dover Street, Piccadilly.

CASE

JOhn Nelfon, formerly of the parifh of Maplebeck, in the county of Notting-ham, mariner, had, for fifteen years, labour-ed under a dreadful complication of the fcur-vy and venereal difeafe. He acquired the former in his youth, by the almoft conftant ufe of falt provifions at fea, and contracted the latter when about twenty years of age. Time had increafed the fymptoms to fo vio-lent a degree that even life itfelf was become a burthen. He had continual running fores in both his legs, a fiftula in perineo, which had penetrated the urinary paffage, fo that he voided as much water through the wound as by the urethra. In this deplorable con-dition he began a courfe of the fyrup, which in about two months removed all his com-plaints, and in lefs than four, was radically cured and enjoyed fince a better ftate of health than he ever could remember in any one pe-riod of his life.

C A S E

CASE IV.

MR. William Painter, of Church Street, St. Anne's, after being for a confiderable time afflicted with a moft inveterate and confirmed venereal diforder, attended with a chancre on the glans penis, of a moft enormous fize, which had eat quite through into the urethra; with a variety of other dreadful fymptoms, excruciating nocturnal pains in the head and limbs, venereal blotches, and, from repeated mercurial courfes, quite emaciated. In this fituation he regularly took a courfe of the Vegetable Syrup; and in one month was perfectly cured.

CASE V.

ALexander Nevie, living with Mr. Collier, a baker, in Bifhopfgate Street Without, was cured by taking four bottles

only

only, of a confirmed lues, being not only afflicted with a moft violent phimofis, but had all the inguinal glands fchirrous, ulcers in the urethra, attended with a pocky hectic, and fo delicate a conftitution, that not the moft gentle dofe of any mercurial preparation that could be given, but what inftantly flung him into the flux, and brought on variety of diforders ; he was feveral times attempted to be cured in the common way, which was ab-folutely impoffible : in this fituation he took the Original Syrup, and in three weeks was perfectly cured.

C A S E VI.

A Gentleman, whofe name I have liberty to mention, was for three years afflicted with ftrictures and ulcerations in the urinary paffage, the effects of an ill cured gonorrhoea, for which he had tried almoft every remedy

extant,

extant, but never received the leaft benefit. In this fituation he took fix bottles of the Vegetable Syrup with unlooked-for fuccefs; and has gratitude enough to declare, that e-very future happy moment of his life dates its origin from the remedy he fo fortunately experienced.

CASE VII.

To Meffrs. SAFFORY *and* MERCIER.

HAving lately received a cure by taking your Vegetable Syrup, which from the nature of my complaints I totally defpaired of, I am induced to give you this public affu-rance of my being indebted, perhaps for life and conftitution, to your invaluable medicine; a moft obftinate fcorbutic complaint had at-tended me for a confiderable time, for which I had repeatedly taken the advice of fome of the moft eminent of the faculty in vain; it

grew

grew at laft fo exceeding bad as to affect one of my eyes, which I was daily in fear of lofing, attended with a total lofs of appetite, and exceffive pains in all my limbs; being recommended to the Syrup by a friend of mine, who had received a remarkable cure, I was induced to try it, which I did with the moft pleafing effect; being perfectly and radically cured of all my diforders by taking five bottles only.

I remain, gentlemen,

Your moft grateful humble fervant,

JOHN BAXTER,

Of the Ladies' Coterie, Albermarle-Street.

CASE VIII.

JAmes Spence, Efq. at No. 12, Caftle Street, near Barner Street, Oxford Road, was, for many years, afflicted with a violent fcurvy, which affected both his legs, as to occafion

cafion continual eruptions, and ulcerations of a moft malignant nature ; after trying many medicines in vain, was radically cured by taking five bottles of the Original Vegetable Syrup. He now remains in perfect health, and is willing, for the general good of mankind, to fatisfy any perfon of the above truth.

CASE IX.

MR. Lever, at No. 6, George Yard, Lombard Street, was afflicted with a furfeit and a violent fcurvy, his face and whole body being full of red fpots and pimples, of which he was radically cured by taking five bottles of the Original Vegetable Syrup.

CASE X.

MR. John Cruchet, of Weft Street, St. Anne's, Soho, was for many years afflicted with a virulent fcurvy, both his legs were

were swelled and full of lived pimples, which occasioned so great a lameness, that it was with great difficulty he could walk, after taking four bottles of the Original Vegetable Syrup was radically cured.

C A S E XI.

To Messrs. SAFFORY *and* MERCIER.

GENTLEMEN,

INduced by a motive of humanity, I request my case may be published, the peculiar misfortunes of which bore so heavy on me that life became a burthen. I was afflicted with the most violent scurvy, my whole body almost one universal scale, particularly my arms and legs, which was pronounced to be a leprosy. I had been an outpatient at St. George's hospital, but met with no relief, the excruciating pains constantly depriving me of rest for several months. I had most of my joints affected with painful and indurated swellings, particularly my right

right ancle, and my knee contracted, which occasioned such a lameness that it was with the greatest difficulty I could stand; I was also excessive weak and in a decline, from a hectic fever, together with a total loss of appetite. The first bottle of the Original Vegetable Syrup greatly abated my pains, five finished my cure in the space of six weeks, and restored me to my former health and strength; which to this moment I enjoy, to the surprise of several people of credit and reputation, who will attest the same. In gratitude to you, and for the good of mankind, I request you to publish my case.

I am, Gentlemen,

Your most obedient humble servant,

ROBERT HUTTON,

At the Penny-Post Office, Lambeth-Marsh, opposite Mount Row.

Witnesses to the cure.

J. WILLIS, Master of the Thatch'd House Tavern, St James's Street.

R. SUTTON, Master of the Ladies' Coterie, Albermarle Street.

CASE

GENTLEMEN,

GRatitude to you, and feeling for my fellow creatures, are the motives of this letter, which I requeſt you will publiſh in juſtice to your moſt excellent Vegetable Syrup, and in order to let the world know the great virtues of that admirable medicine.

I had been for a long time afflicted with a violent ſcurvy, complicated with a general rheumatiſm of my whole body, which rendered me ſo feeble and helpleſs that I could not turn myſelf in bed without aſſiſtance. I laid in this melancholy ſituation for upwards of three months, with the additional agony of pains from head to foot ; a waſting hectic, and a total loſs of appetite to ſtruggle with.

Two

Two eminent perfons of the faculty conftantly attended me, without being able to procure
me the leaft relief; or alleviate my diforder.
Nature being quite exhaufted, I defpaired of
recovery, when Mr. Hutton, (at the penny
poft office, Lambeth Marfh, oppofite Mount
Row) an old patient of yours, and one who
had already experienced the certain efficacy
of your moft admirable remedy, recommended
me to you.

After taking the firft bottle, I found my
ftrength fo amazingly reftored, that I could
not only turn in my bed without help, but
even walk my room. In fhort, I recovered
fo faft that in about a fortnight I was able to
go out, and when I had taken fix. bottles,
was (by the grace of God) radically cured of
all my complaints, and found myfelf once
more in poffeffion of thofe ineftimable bleffings,
health and tranquility. The truth of the
above is very well known to all the neigh-

L   bours,

bours, who are ready to atteſt the ſame to any enquirer.

I am,

With the warmeſt ſenſe of gratitude,

Gentlemen,

Your moſt obliged humble ſervant,

T H O M A S   H O W E L S,

Farrier, Lambeth Marſh.

Witneſſes to the cure,
JOS. HOPKINS, Surgeon,
No. 85, Frith Street, Soho.
ROBERT HUTTON,
Lambeth Marſh.

C A S E   XIII.

MR. Beaumgartner, merchant, of Poor Jewry Lane, Aldgate, reſiding at Chelſea, was, for a number of years, afflicted with a moſt virulent ſcurvy, which had occaſioned one univerſal ſcurf over his head, which conſtantly diſcharged a large quantity of peccant

foetid

fœtid matter : there was alfo upon his forehead a lump as big as a pigeon's egg ; feveral of the joints were enchylofed, particularly
the left arm, which he had nearly loft the
ufe of. At the fame time he was afflicted
with fevere complaints, univerfal laffitude,
and a want of appetite; together with a conftant and moft excruciating rheumatic pains
in all his limbs. In this deplorable ftate, he
applied to M. de Velnos, when he was laft
in town, who gave his Syrup with the greateft fuccefs: the fourth bottle perfected the
cure, except a little lamenefs, which was removed after M. de Velnos's departure, by
adminiftering of a fifth bottle, the patient
was radically cured; and now remains in perfect health.

CASE XIV.

MR. John Man of Cold Bath Fields,
was moft feverely affected with a complication of the fcorbutic rheumatifm; he had
loft

loſt the uſe of one arm for three years before, which limb never had the appearance of any eruptions, the other was conſtantly covered with large puſtules like the ſmall pox. Every application both external and internal had failed. A phyſician adviſed him to try the Original Syrup, which he did with ſurpriſing effect; the firſt bottle that he took apparently diſagreed with him, and made him rather hotter than before; the ſecond threw him at times into a violent itching of the blood, and ſeveral ſpots appeared on the lame arm, which in a fortnight was covered with a thick raſh, he then began to feel a free circulation in the limb, and by perſeverance in the uſe of the Original Syrup, he perfected a cure, and has the perfect uſe of all his limbs.

C A S E XV.

JOhn Hanſon, a mariner, who from living on ſalt proviſions had acquired the ſcurvy in ſo violent a degree as to be covered over

entirely

entirely, body and legs; as a leopard with an univerſal white ſcale that would ſcrape off like chalk, a continual nauſeous and vomiting, and could digeſt nothing the leaſt ſolid; ſix bottles of the Syrup perfectly cured him, and he now remains well and in perfect health.

C A S E XVI.

M R. James Sprags of Spitalfields was afflicted with ſcorbutic eruptions over moſt parts of his body, which conſtantly diſcharged an iſchorus matter prodigiouſly offenſive, and of ſo hot a nature as to excoriate the ſkin wherever it touched, his general conſtitution was much emaciated, dwindled to a mere ſkeleton from exceſs of pain; was relieved by a few bottles of the Original Syrup, and perfectly cured in ten weeks.

C A S E

CASE XVII.

MR. Alexander Donald, of Margaret
ſtreet, Cavendiſh ſquare, after being
affected in the moſt violent manner for ſix
years with the ſcorbutic humour ſo interwo-
ven in his conſtitution as to occaſion a caries
of the ſhin bone, which never could be
brought to ſeparation, tho' he had tried fifty
different remedies, and had been ſalivated
twice; by taking five bottles of the Original
Vegetable Syrup the bone exfoliated, in two
months he was perfectly cured, the wound
healed, and his conſtitution reſtored to its
priſtine health.

CASE XVIII.

MR Mathews, formerly of Chancery
Lane, was for fourteen years, afflicted
with the ſcurvy every ſpring and fall, he was
covered

covered with running fores, which, by the ufe of falt bathing, was a little kept under; he was at Margate at the time he applied for the Original Syrup, at which place he took five bottles, and was radically cured, that is two years ago, and he has not had the leaft return fince.

CASE XIX.

J. Watts, a labourer, afflicted in the moft terrible manner with fcorbutic erup-tions all over his body, and a wound upon one of his legs, for which he had been in four different hofpitals without the leaft benefit, five bottles of the Original Syrup perfectly cured him.

CASE

CASE XX.

MRS. Martin, of Horden, was for several years in the most alarming situation, from a violent swelling of the abdomen, and a prodigious indurated lump as big as a cricket ball, of the right side; she had generally a yellow complection, bordering on the jaundice, universally obstructed with scorbutic eruptions and pimples at times, her disorder was pronounced a bad liver, and supposed to be fatal, but fortunately her physician who attended her advised Velnos's Original Syrup, which by taking regularly with the most minute attention for three months, she became gradually restored to a state of perfect health, which she enjoys to this hour.

CASE XXI.

ANdrew Moss, at Bow, was for upwards of seven years afflicted with a scorbutic nervous headach, so as to prevent his

working

working at his trade, which was that of a carpenter, he was frequently taken in an instant with such excruciating pains in his forehead as to occasion partly the loss of sight; he had on one arm some scaly white pustules, which at times discharged, then he was free from pains in his head; he took five bottles only of Velnos's Original Vegetable Syrup, which perfectly and radically cured him.

<hr />

CASE XXII.

MR. Marchmont, a foreigner, residing in Mark Lane, had from his infancy been afflicted with a leprous scorbutic, he was at times one universal scale; six bottles of the Original Syrup, and a great attention to regularity of diet perfectly cured him.

CASE

CASE XXIII.

MR. Goodwin, of Lambeth, had, by requeſt of ſome of the faculty, been ſalivated for ſome ſwellings that were of an alarming nature, the glands under the armpits, in the throat and groins were prodigiouſly indurated, but did not give the leaſt way to the ſtrongeſt mercurial courſes that could be uſed, both external and internal; it was judged ſcrophulous, he was adviſed to try the Original Syrup, which he did with the deſired effect, in leſs than two months he was perfectly cured.

CASE XXIV.

MR. Munroe, reſident at Brighthelmſtone, was for fourteen years afflicted with ſcorbutic eruptions in his face and neck,

for

for which, he had applied to feveral of the faculty with no fuccefs; his head, face and fhoulders being one univerfal fcab, five or fix bottles of the Original Syrup made him perfectly well.

CASE XXV.

MR. Warren, a gentleman whofe afflictions in point of illnefs was incredible, never knowing a fingle day's eafe or health for many years; he was afflicted by fcorbutic obftructions, hereditary, his whole family being eat up with the fcurvy, his feet, legs and arms were one fore, at times difcharging plentifully, at others quite dry, with a white fcale, his face covered with red pimples; he was alfo affected with a conftant nervous head-ach, and had the rheumatifm in every limb. In this fituation, exhaufted by repeated trials of almoft every medicine extant,

tant, he took a regular courfe of the Original Syrup, which in two months cured him.

CASE XXVI.

PAul Martin, of Birmingham, coachman to a gentleman of that place, cured of a confirmed lues, after trying different preparations of mercury in vain, by taking a few bottles of the Original Vegetable Syrup.

CASE XXVII.

MR. John Bradwick, of Plummer's Court, Holborn, cured of an inveterate lues. See his cafe, written by himfelf, which cannot with decency be printed.

CASE

CASE XXVIII.

MR. Raimond, valet de chambre to a nobleman, cured of an inveterate scurvy in his hands and feet, complicated with a certain disorder, by taking six bottles of the Original Vegetable Syrup.

When it is remembered how few persons choose to have their names publicly mentioned, especially in venereal cases, no apology will perhaps be expected for inserting the following cures performed on anonymous patients: but have obtained free liberty to refer any candid enquirer to those patients, who will very readily give them abundant satisfaction.

CASE XXIX.

MR. and Mrs. G. of Crown Street, Soho, were cured of a confirmed and inveterate venereal complaint, of twenty-one years standing,

standing, which diforder being fo complicated with the fcurvy, that it baffled every medicine, adminiftered by feveral perfons of the firft confequence in their profeffion; they had alfo took without the leaft effect thirty bottles of Burrows's Syrup with improvements. Six only, of the Original Syrup, adminiftered to each, radically cured and reftored them both to their perfect health and ftrength. The particulars of their diforder, and the above cafe is written in the hufband's hand writing, with permiffion to fhew it to any candid enquirer, and is to be feen by applying to Dr. Mercier, at No. 39, Frith Street, Soho.

CASE XXX.

C. F. Efq. of Devon, by unfortunately loading his conftitution with mercurials, together with a natural fcorbutic habit, had by the violence of the diforder loft

the

the fight of one eye, the other could fcarce diftinguifh large objects, his gums were entirely eat away, feveral of the teeth were carious, and a fmall portion of the lower jaw difeafed, generally dibilitated, with night fweats, and a bed hectic cough; in this fituation Meffrs. Saffory and Mercier were fent for to fee him, they advifed his being brought to London, and placed under their care, when by a careful adminiftration of the Original Vegetable Syrup, they had the pleafure to return him to the longing arms of a large and excellent family, in about three months, perfectly well; a large portion of the jaw feparated, feveral of the teeth were extracted, and after taking only three bottles, he could fleep half the night with perfect eafe, loft his cough, and by degrees every unfortunate fymptom vanifhed.

CASE

C A S E XXXI.

T. W. Union Stairs, Wapping, from a constant use of salt provisions, had the scurvy settled in the point of his right arm, which became quite scrophulous, a large wound broke out near the tip of the elbow, from whence a large discharge issued for several months; in less than a fortnight, by taking the Original Vegetable Syrup, the wound bore a different aspect, the discharge lessened, and in six weeks he was perfectly well.

C A S E XXXII.

L. B. of Onger, in Essex, a grazier, laboured under a complication of disorders, particularly a deafness, occasioned, as was pronounced by a scorbutic humour in the blood, which shewed itself even in his infant state;

state; he had for several years a foetid discharge of wax and matter from both ears, and often extremely deaf; at times the glands of the neck enlarged, and once suppurated; he had tried innumerable remedies, and had applied in vain to two persons famous in disorders of the ears: a gentleman in the neighbourhood who had received a cure by the Syrup, recommended it to the above person, whom, by taking seven bottles, in the space of nine weeks became perfectly well, and has remained so ever since.

CASE XXXIII.

H G. of the island of Guernsey, was confined to his room seven months with the rheumatism, complicated with some other disorders, particularly the scurvy, to so violent degree that he was constantly obliged to be dressed by a surgeon twice a day, for

N

three

three months, having running fores in both legs and arms, moft of his joints were abfolutely echylofed, and by the ufe of crutches only he could crawl acrofs the room, he was then attended by a phyfical perfon in the ifland, who had adminiftered the Original Syrup with furprifing effect; he put this perfon under a ftrict courfe of it, which in lefs than four months, perfectly reftored him to the free ufe of all his limbs, and a perfect reftoration of health.

CASE XXXIV.

L. U. of the ifland of Grenades, was cured by taking four bottles only of the Vegetable Syrup, after every other method had failed, of a virulent fharp fcorbutic humour, that excoriated every other part it fell on; in this cafe, it was remarkable that he had taken two bottles only, when the ifchor that dif-

charged

charged from the fcorbutic eruptions had loft its ufual heat; for, before, the very rheum that came from his eyes, caufed the fkin to come off from the face; in one month he was perfectly cured.

CASE XXXV.

H. G. recommended as a pauper, by the parifh officers, was cured of an inveterate fcurvy which had broke out in red pimples all over his body, with fo violent an itching as to occafion him to tear the fkin off in fuch a manner, that he could not turn from one fide to the other. He took five or fix bottles of Velnos's Original Syrup, and was perfectly cured, and remains without the leaft return of his diforder.

CASE

CASE XXXVI.

ANother perſon, recommended by the ſame officers, received a cure in a ſimilar caſe, except this laſt was complicated, he having the jaundice and dropſy at the ſame time, but as it all aroſe from one grand cauſe, Obſtructions, and that ſcorbutic, the Syrup had a happy effect, and he was well in a ſhort time.

CASE XXXVII.

W. T. recommended by Mr. Thomſon of Cheſhire, had from his infancy, occaſioned by a ſurfeit, one of the moſt virulent ſcorbutic complaints that could poſſibly be, attended with ſeveral aggravating circumſtances, ſuch as a general debility, night ſweats, and ſwellings in the legs; he took only four bottles, of the Syrup, and was, and now remains in perfect health.

CASE

C A S E XXXVIII.

A Gentleman who keeps a lottery office, being afflicted with an inveterate fcurvy, after fpending five hundred pounds in various medicines and advice in vain, was radically cured by taking feven or eight bottles of Vel-nos's Original Vegetable Syrup.

C A S E XXXIX.

A Gentleman refiding in Holborn, was af-flicted with feveral cutaneous irruptions in the face, occafioned by a fharp fcorbutic humour in the blood, was perfectly cured by taking fix bottles of the Original Vegetable Syrup.

C A S E

CASE XL.

A Gentleman of Marſham Street, Weſt-minſter, cured of carnoſities and ſtric-tures in the urethra, with excoriations on the præputium of twenty years ſtanding, by the uſe of a few medicated bougies and ſeven bot-tles of our ſyrup, he was radically cured.

CASE XLI.

AN officer cured of an inveterate lues, af-ter trying the ſtrongeſt mercurial pre-parations in vain, was radically cured by taking ſeven bottles of the Original Vegetable Syrup.

CASE XLII.

AN elderly lady, cured of an inveterate ſcorbutic complaint and pimpled face, occaſioned by a ſurfeit, by taking five bottles of the Original Vegetable Syrup.
CASE

CASE XLIII.

A Valet de chambre to a nobleman had almoſt loſt the uſe of his left arm, by the uſe of mercury, and was full of pains in the joints, eſpecially his ancles; he was radically cured by taking ſix bottles of the Original Vegetable Syrup.

CASE XLIV.

THe daughter of a gentleman had ſeveral ulcers in her legs, one in particular on the ſide of the tibia, very deep and of a malignant nature, the ſkin of her legs was like the ſcale of a fiſh, ſhe was radically cured by taking at two different times a courſe of the Syrup. She was recommended to waſh the ſores with the Syrup, and in leſs than three months, was perfectly cured.

CASE

CASE XLV.

A Young lady of Southampton, afflicted with an inveterate fcorbutic humour and pimpled face, was perfectly cured by taking five bottles of the Original Vegetable Syrup.

CASE XLVI.

A Country gentleman had feveral irruptions in his face, arms and legs, occafioned by a furfeit, complicated with the fcurvy, was radically cured by taking fix bottles of Velnos's Original Vegetable Syrup.

CASE XLVII.

To Meffrs. SAFFORY *and* MERCIER,

GENTLEMEN,

I Should think myfelf wanting of gratitude and even of humanity, was I to delay any longer to acquaint you of the extraordinary

cure

cure I received by the use of your Vegetable Syrup. I had been afflicted for several years with a most malignant scrofulous and scorbutic humour all over my body and face, especially in my head, arms and legs, which constantly discharged an ischorus matter very offensive, and of so hot a nature as to excoriate the skin of my face wherever it touched, my constitution very much debilitated, and dwindled to a mere skeleton from excess of pain. After trying several medicines in vain, which did me more harm than good, a gentleman of this place recommended me to your Vegetable Syrup, having performed a surprising cure on one of his friends; I applied to you, and after the second bottle I found myself stronger, having a better appetite; the swellings in my neck disappeared gradually, the discharge of my sores were less offensive, and in two months time was radically cured of all my complaints, to the great joy of all my acquaintance, and have enjoyed ever since a perfect state of health. In gratitude to you,

O   and

and for the good of thofe afflicted as I was, I requeft you to publifh my cafe.

I am, Gentlemen,

Your moft obliged humble fervant,

J O H N L A N E,

Mafter of the Angel Inn and

Livery ftable, Birmingham.

Birmingham, July 25, 1775.

Witneffes to the cure,
THO. WARREN.
P. F. BOURGEOIS.

C A S E XLVIII.

To Meffrs. SAFFORY *and* MERCIER.

GENTLEMEN,

THe extraordinary cure I have lately received by a courfe of your Velnos's Vegetable Syrup, binds me in gratitude to make known to the public, the wonderful effects of that moft falutary remedy. At the time, I begun the Syrup, I was afflicted with the moft melancholy and fevere venereal fymp-

toms

toms that could poffibly be; ulcerations of
the worft nature in the throat and palate,
almoft all I drank came through the nofe,
the bones of which were confiderably affected.
I had conftant excruciating pains in all my
limbs, one of my knees was enchylofed, the
ancles as well as fhin bones confiderably enlarged. I had been falivated in the Lock
Hofpital, fince which, I have been under the
hands of many of the faculty, and of the firft
confequence in their profeffion; tired of life,
and loft to defpair, I was advifed to take
your Syrup, by a perfon receiving a cure in a
fimilar cafe. I am now, by taking fix bottles
of it, in perfect health, radically cured of all
my diforders, and a living witnefs of the great
efficacy of the Original Vegetable Syrup; in
gratitude for benefits received, I requeft my
cafe to be publifhed, and am refpectfully,

  Gentleman,

   Your ever obliged humble fervant,
    GEORGE OLIVER.

Long-Acre, November 23, 1775.

       CASE

Extract of a Letter *from* Paris, *Feb.* 3, 1776.
(Tranflated from the French.)

To Dr. MERCIER.

S I R,

I Have the pleafure to acquaint you that
Mr. De Velnos's Syrup hath performed
here lately a wonderful cure. The fecretary
of the Venetian Ambaffador, after a courfe
of mercury had failed, had his throat ul-
cerated all over, and could take no food but
by the means of a fyringe, fo that there was
hardly paffage to convey through that inftru-
ment a little broth, and by taking a few bottles of the Syrup (adminiftered by Dr. Mi-
quet,) he was radically cured, to the great
aftonifhment of the faculty and advocates of
mercury.

I am, Sir,

Your moft obedient fervant,

LE CLERC.

CASE L.

To Meſſrs. SAFFORY and MERCIER.

GENTLEMEN,

HAving received a perfect cure of a bad complication of a nervous and ſcorbutic diſorder which had afflicted me for ſeveral years, with the moſt violent pains in my head, much emaciated, and reduced by low ſpirits. I fortunately took ſeven bottles of your Original Velnos's Vegetable Syrup, which has reſtored me to my proper health, after trying various medicines and variety of advice in vain. I am deſirous of making this known, and therefore give you leave to publiſh it.

I am, Gentlemen,
Your moſt obliged humble ſervant,
ELIZABETH RYLAND.

No. 25, Bennet Street, Weſtminſter, July 10th 1776.

CASE

CASE LI.

To Meſſrs. SAFFORY *and* MERCIER.

GENTLEMEN,

HAving been for upwards of ſix years, afflicted with a moſt violent ſcorbutic
humour in my arms and legs, complicated
with the rheumatiſm, which almoſt totally
deprived me of the uſe of my limbs, ſo as to
diſenable me from carrying on my uſual buſi-
neſs, and, by taking five bottles of your Velnos's Original Vegetable Syrup, am perfectly
cured : After the ſecond bottle I was able to
walk ſeveral miles upon the ſtretch, which I
had not been able to perform for theſe ſeveral
years paſt, and by the grace of God and the
aſſiſtance of your Vegetable Syrup, I enjoy at
preſent a perfect ſhare of health. In juſtice
to your moſt excellent medicine, and for the
benefit

benefit of thofe afflicted as I was, I fend you this extraordinary cafe, with my confent to publifh it.

I am, GENTLEMEN,

Your moft obliged humble fervant,

FRED. POTTERAT.

No. 13, Bow Lane,
July 10, 1776.

CASE LII.

To Meſſrs. SAFFORY *and* MERCIER.

GENTLEMEN, .

I Return you my fincere thanks for the great benefit I have received, in taking your Vegetable Syrup, by the excellence of which I have been cured of a continual pain, which occured from having caught cold in fifhing, &c. this pain was fo violent, that I could not reft in the night, of which I am, thank God, radically cured, and am very

willing

willing to give the teſtimony of the above to any inquirer.

I am, Gentlemen,

Your obedient humble ſervant,

CHARLES SAVARY.

Coleman Street, No. 56,
Auguſt 13, 1776.

F I N I S.

E R R A T U M.

At the End of Caſe XXVII. *read,* " This caſe is writ-
" ten by the Patient, and is to be ſeen by enquiring
" at No. 39, FRITH STREET, SOHO."